How Animals Evolved

HOW INSECTS EVOLVED

DiscoverRoo
An Imprint of Pop!
popbooksonline.com

by Grace Hansen

WELCOME TO DiscoverRoo!

This book is filled with videos, puzzles, games, and more! Scan the QR codes* while you read, or visit the website below to make this book pop.

popbooksonline.com/ev-insects

abdobooks.com

Published by Pop!, a division of ABDO, PO Box 398166, Minneapolis, Minnesota 55439.

Printed in the United States of America, North Mankato, Minnesota.

102023
012024

THIS BOOK CONTAINS RECYCLED MATERIALS

Cover Photo: Shutterstock
Interior Photos: Shutterstock; Getty Images; Science Source
Editor: Elizabeth Andrews
Series Designer: Laura Graphenteen

Library of Congress Control Number: 2023939057

Publisher's Cataloging-in-Publication Data

Names: Hansen, Grace, author.
Title: How insects evolved / by Grace Hansen
Description: Minneapolis, Minnesota : Pop!, 2024 | Series: How animals evolved | Includes online resources and index
Identifiers: ISBN 9781098245443 (lib. bdg.) | ISBN 9781098246006 (ebook)
Subjects: LCSH: Evolution--Juvenile literature. | Insects--Juvenile literature. | Insects--Evolution--Juvenile literature. | Biology--Juvenile literature.
Classification: DDC 591.38--dc23

*Scanning QR codes requires a web-enabled smart device with a QR code reader app and a camera.

TABLE OF CONTENTS

CHAPTER 1

WHAT IS EVOLUTION?

Evolution is the scientific **theory** that all life on earth has a common ancestor that lived billions of years ago. Living things **developed** from earlier versions of that ancestor. The changes that happened took millions of years.

WATCH A VIDEO HERE!

Ladybugs have special features that help them survive. Their brightly colored bodies warn predators to stay away.

DID YOU KNOW?

Almost all scientists believe in evolution because it is testable.

***Stromatolites* (in water) are some of the oldest fossils on Earth, dating back more than three billion years when the planet was young.**

Evolution happens because of natural selection. This process occurs when one individual of a **species** has different **traits** from others. Sometimes the differences make surviving easier for

the individual. As they survive, they pass on their traits to their young. Over time, more and more individuals in the species are born with the new and helpful trait until it is normal.

Evolution is how fish eventually moved onto land. It is how birds began to fly and mammals started giving birth to live young. It is how humans began walking on two feet.

Archaeopteryx, a dinosaur that lived 150 million years ago (MYA), likely ate dragonflies.

CHAPTER 2

WHAT ARE INSECTS?

Insects are the most varied of all **invertebrates**. They are essential to life on Earth. They **pollinate** fruit trees. They break down dead plants and animals to make soil. Many other animals rely on insects for food.

LEARN MORE HERE!

A dragonfly's wings allow the insect to move in all directions.

The unique lantern bug can be found in parts of Southeast Asia.

There are more than one million insect **species**. They live throughout the planet in many different habitats. Insects range from beautiful dragonflies and butterflies to ants and fruit flies. However, they all share some similar features.

Butterflies spread pollen from flower to flower as they feed on nectar.

All insects have an **exoskeleton**, three pairs of legs, one pair of antennae, and a segmented body. The body is made up of the head, thorax, and abdomen. Most adult insects have wings.

Bees gather nectar to bring back to their hive. They use it to make honey.

MAIN INSECT BODY PARTS

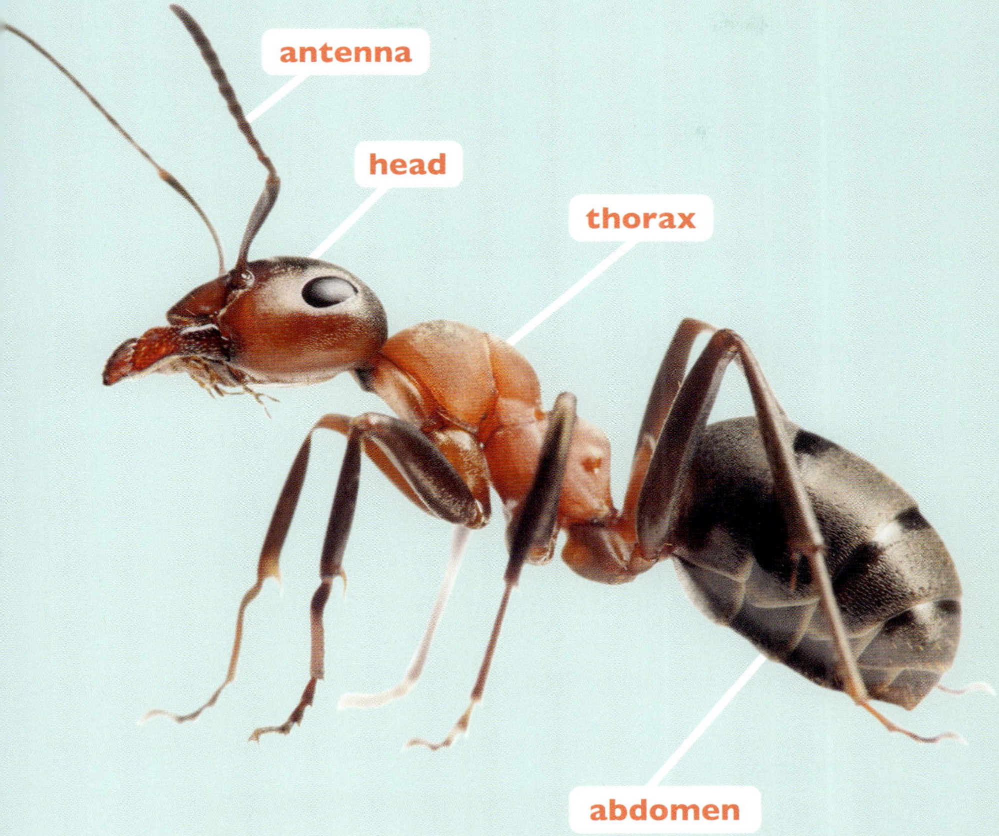

An insect's head contains the mouthparts, eyes, and antennae.
The legs and wings are attached to the thorax.
The abdomen contains important organs.

CHAPTER 3

BEFORE BUGS

Half a billion years ago there were no insects. Earth looked very different. On land, there were no animals or plants. Nothing flew in the sky. However, the ocean was full of life.

EXPLORE LINKS HERE!

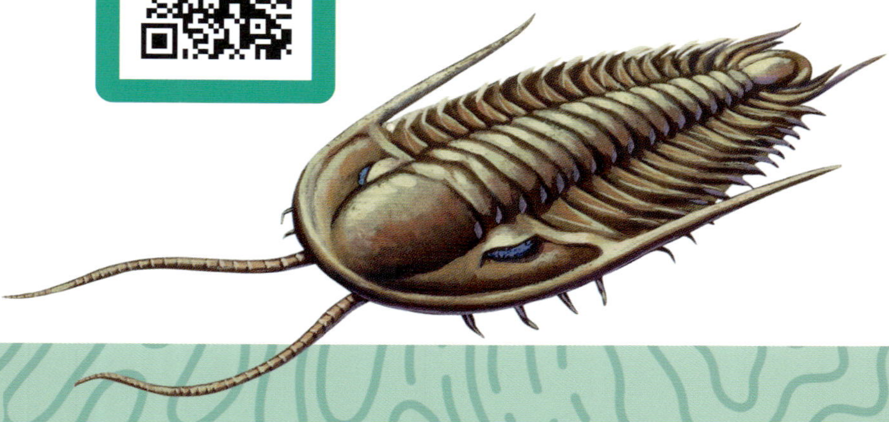

Amber naturally preserves plant matter, insects, spiders, and other small creatures.

Simple plants called algae floated on the ocean surface. This provided food for the marine life below. Strange creatures

Trilobites lived around 542 MYA. They are related to crustaceans and insects.

called trilobites looked like giant pill bugs. They lurked on the muddy seafloor in shallow water.

At some point, plants made their way onto land. These early plants provided homes and food for the first creatures that moved from water to land. Springtails may have been some of the earliest land dwellers. These small, wingless arthropods look much like insects.

SPRINGTAILS

Springtails got their name because of the taillike body part they used for leaping. They're so similar to insects that scientists used to believe they were closely related. However, further studies put them in their own order.

DID YOU KNOW?

The oldest insect fossil is thought to be 400 million years old. It is an insect named *Rhyniognatha hirsti*.

Scientists know from fossil and other evidence that all the insects living today came from the same group of animals. A study in 2014 revealed that insects descended from crustaceans, another branch of arthropods. Insects are related to crustaceans known as remipedes. Remipedes are blind, long-bodied cave dwellers that still exist today.

Crustaceans are invertebrates with a hard exoskeleton. Crabs, lobsters, and shrimp are all types of crustaceans.

CHAPTER 4

STAGES OF INSECT EVOLUTION

Insects have evolved to live all over the world. They have different features, habitats, and **predators**. Today, insects account for 80% of the world's **species**. Certain events, called stages, helped insects take over the world.

COMPLETE AN ACTIVITY HERE!

Butterflies and moths, such as the small emperor moth, begin life as caterpillars.

In the first stage of insect evolution, the first wingless insects appeared. One type of early insect was the bristletail. Bristletails are considered the most ancient insects. They still live around the world today.

Some caterpillars have special ways to defend themselves.

The second stage was when some insects evolved wings that can't fold, such as those on dragonflies. In the third stage, other insects evolved wings that could fold, such as on cockroaches. Wings helped insects move easily from place to place. There were also fewer **predators** in the sky.

WONDERFUL WINGS

With wings came more evolutionary changes. Some insects use their wings as **camouflage**. Others use their wings for protection, to make sound, or control body temperature.

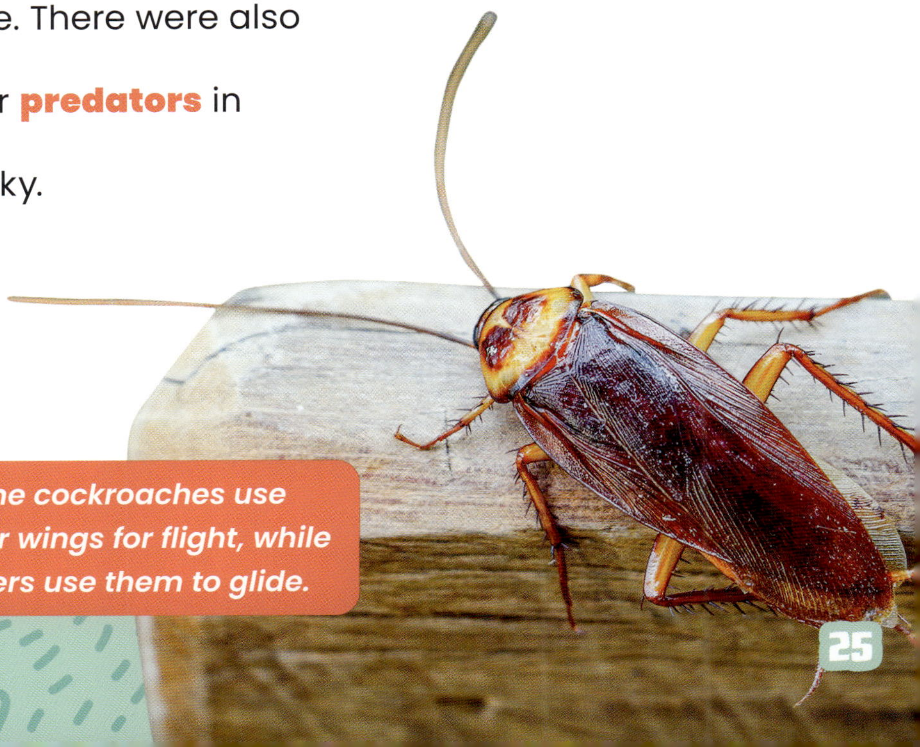

Some cockroaches use their wings for flight, while others use them to glide.

BUTTERFLY METAMORPHOSIS

The fourth stage was the evolution of complete metamorphosis. One example of this is the transformation of caterpillars to butterflies. Insects with significantly different life stages face less competition for food and habitat. However, not all insects use metamorphosis to grow.

Butterflies go through big changes during their life cycle.

Some simply grow in size and shed their **exoskeleton**.

Finally, social groups **developed**. For example, ants or bees live and work together in groups. Both ants and bees work to build shelter, find food, and care for young.

Insects are very important to the global **ecosystem**. Recently, crickets have been used as food. Insects provide a protein source that is better for the Earth than meat from mammals. Scientists are still discovering and studying fossils of ancient insects. The story of insect evolution is still being written by scientists today.

DID YOU KNOW?

The stages of insect evolution happened over millions or even tens of millions of years.

Some insects, such as cicadas, shed their skin as they grow.

MAKING CONNECTIONS

TEXT-TO-SELF

What part of the evolution of insects surprised you the most? Please explain your answer.

TEXT-TO-TEXT

Have you read any other books about animal evolution? What did the evolution of those animals have in common with insect evolution?

TEXT-TO-WORLD

There are thousands of species of insects in the world. With the help of an adult, look up information about your favorite insect. Write a short paragraph explaining why it is your favorite.

GLOSSARY

camouflage — a way of hiding something by covering or coloring it so that it looks like its surroundings.

develop — to grow over time.

ecosystem — a community of organisms and their surroundings.

exoskeleton — an external supporting structure such as the shell of a crustacean.

invertebrate — an animal that does not have a backbone or skeleton inside its body.

pollinate — to move or carry pollen to a plant, causing the seeds to be fertilized.

predator — an animal that lives by hunting and eating other animals.

species — a group of living things that are very much alike.

theory — an explanation for how things work and why things happen.

trait — a quality or feature of something.

INDEX

DiscoverRoo!
ONLINE RESOURCES

This book is filled with videos, puzzles, games, and more! Scan the QR codes* while you read, or visit the website below to make this book pop.

popbooksonline.com/ev-insects

*Scanning QR codes requires a web-enabled smart device with a QR code reader app and a camera.